Alejandro Nesci
Ricardo Baldizon
Carlos Mendoza

Solid Foundations

Alejandro Nesci
Ricardo Baldizon
Carlos Mendoza

Solid Foundations

The intersection of engineering and management in infrastructure.

ScienciaScripts

Imprint
Any brand names and product names mentioned in this book are subject to trademark, brand or patent protection and are trademarks or registered trademarks of their respective holders. The use of brand names, product names, common names, trade names, product descriptions etc. even without a particular marking in this work is in no way to be construed to mean that such names may be regarded as unrestricted in respect of trademark and brand protection legislation and could thus be used by anyone.

Cover image: www.ingimage.com

This book is a translation from the original published under ISBN 978-613-9-40487-2.

Publisher:
Sciencia Scripts
is a trademark of
Dodo Books Indian Ocean Ltd. and OmniScriptum S.R.L publishing group

120 High Road, East Finchley, London, N2 9ED, United Kingdom
Str. Armeneasca 28/1, office 1, Chisinau MD-2012, Republic of Moldova, Europe
Managing Directors: Ieva Konstantinova, Victoria Ursu
info@omniscriptum.com

Printed at: see last page
ISBN: 978-620-8-50906-4

SOLID FOUNDATIONS: THE INTERSECTION OF ENGINEERING AND MANAGEMENT IN INFRASTRUCTURE

AUTHORS: ALEJANDRO ENRIQUE NESCI MONTALVAN RICARDO

JOSÉ BALDIZÓN LÓPEZ

CARLOS ALEXANDER MENDOZA JACOMINO

TABLE OF CONTENTS

CHAPTER 1
THE INTERSECTION OF ENGINEERING AND MANAGEMENT IN INFRASTRUCTURE

1.1 Introduction

Infrastructure construction is a fundamental discipline for the development of any modern society, as it lays the foundations for the creation of essential systems that support the daily functioning of cities, communities and regions.
entire cities. From highways and bridges to water distribution systems, electricity, telecommunications networks and hospitals, each infrastructure project represents a multi-dimensional effort that not only responds to immediate needs, but seeks to lay the foundation for long-term growth and sustainability (Project Management Institute, 2021).

The infrastructure development process involves multiple stages: from needs identification and initial planning, through technical design and implementation, to operation and long-term maintenance. At each of these stages, technical precision is crucial, and this is where engineering plays an essential role. Engineering provides the scientific knowledge necessary to design and build safe and efficient structures, ensuring that they meet the standards of safety, functionality and durability that a project of this nature demands (Bent, J. A., & Humphreys, K. K. (2009).

However, the implementation of infrastructure projects goes far beyond the purely engineering scope (Chan, 2014). While engineering ensures that the technical and structural aspects of projects are feasible and meet quality and safety requirements, the true consolidation of these projects is only possible through efficient administrative management. This management encompasses financial planning, cost control, strategic allocation of resources, management of human resources, risk mitigation, compliance with deadlines and constant monitoring of the progress of the works. Without a solid administrative foundation, even the most innovative and effective technical design could face significant operational, sustainability and even legal compliance challenges.

In this context, the integration of management and engineering in infrastructure becomes particularly relevant. Today's infrastructure projects are becoming increasingly complex, costly and ambitious, which requires close and constant collaboration between engineers and engineers in the infrastructure sector. and managers from the planning phase through to project completion. This interdisciplinary collaboration allows each discipline to bring its strengths to bear: engineering ensures technical accuracy, innovation and compliance with structural specifications, while management ensures that every aspect of the project stays within the necessary budgetary, time and regulatory constraints. In an environment where demands for sustainability and social responsibility are constantly growing, this collaboration becomes an imperative, allowing projects to be both technically sound and financially viable and socially responsible (Rodriguez et al., 2019). The importance of the intersection between engineering and management is even more evident in a context where resources are limited and the demands of environmental sustainability and social responsibility are increasing. Nowadays, infrastructure projects must be designed and executed with a long-term sustainability perspective, which means minimising environmental impact, optimising the use of resources and ensuring the durability of constructions over time. Project management in infrastructure is not only concerned with costs, but also with incorporating sustainable approaches and practices in planning and execution, such as using recyclable materials, designing energy-efficient structures and reducing construction waste.

(Bent, J. A., & Humphreys, K. K. (2009).This chapter explores in depth the intersection of engineering and management in infrastructure, highlighting the importance of interdisciplinary collaboration to achieve project goals. It examines the different phases of an infrastructure project and how engineers and managers must work together to overcome technical, economic and environmental challenges. Strategic planning, management of human and material resources, quality control and compliance with local and international regulations are just some of the aspects that require effective integration of both disciplines. Through a detailed analysis of each of these areas, it is intended to demonstrate that the synergy between engineering and management not only facilitates the execution of more efficient and cost-effective projects, but also helps to improve the quality and efficiency of the work. higher quality, but also promotes infrastructure that contributes to societal well-being and economic progress (Ahmed, S., Farooqui, R., & Saqib, M. (2019).

Ultimately, collaboration between engineering and management in infrastructure allows for stronger, more resilient projects that are adapted to the changing needs of communities. Infrastructure projects that effectively integrate both disciplines are able to respond to today's challenges of rapid urbanisation, climate change and resource demands, providing solutions that are both innovative and sustainable. Modern infrastructure must not only meet current needs, but also anticipate future ones, and it is precisely in this long-term vision that the collaboration between engineering and management becomes crucially relevant. (Smith, P. G., & Reinertsen, D. G. (2014).

1.2 Planning and Design of Infrastructure Projects

In the planning and design phase, engineers and managers play complementary roles that are essential to the success of any infrastructure project. This initial stage is critical as it lays the foundation for the development of the project by defining its objectives, scope and requirements as clearly as possible. During this phase, engineers and managers work together to identify the needs and expectations of stakeholders such as project sponsors, affected communities and regulators. Interdisciplinary collaboration in this phase helps ensure that all aspects of the project are considered before moving forward to implementationminimising the risk of unforeseen problems and cost overruns (Kerzner, 2017).Management involvement in the planning phase is key to assessing financial feasibility and conducting a cost-benefit analysis. This analysis makes it possible to determine whether the project can be implemented within the available resources and, if is not feasible, to identify alternatives or adjustments that allow for its implementation. Managers are responsible for projecting the costs of each of the activities that make up the project, from design through maintenance to construction. They also consider the available budget and establish a financial plan that covers all the necessary aspects. This can include everything from seeking additional sources of funding, such as loans or grants, to planning payments and estimating returns on investment for projects that will generate revenue. This early financial analysis helps to ensure that the project is sustainable throughout its life cycle and that resources are managed efficiently (Bent, J. A., & Humphreys, K. K. (2009). Engineers, on the other hand, focus on technical feasibility, a crucial component that defines how the project will be carried out structurally, functionally and in terms of design. They assess ground conditions, available materials, infrastructure needs and the technologies to be employed to ensure that the project is technically feasible.

This work includes the selection of appropriate technologies and the development of preliminary designs to suit the specific site conditions and the technical and functional requirements of the project. In addition, engineers work to identify potential technical challenges, such as soil type, climatic conditions and other factors that could affect the development of the project. Thanks to their specialised knowledge, they can propose technical solutions to achieve the project's objectives safely, efficiently and in compliance with quality standards (Walker & Rowlinson, 2008). Detailed planning at this stage also helps to minimise risks and unforeseen costs, which is critical to the long-term viability and success of the project. A comprehensive plan includes the development of timelines detailing each stage of the project, from site preparation to final delivery, and includes contingencies that may arise. Management is responsible for establishing these schedules in collaboration with the engineering team, ensuring that realistic timelines are considered and resources are allocated appropriately. In addition, risk management strategies are implemented, such as identifying potential problems and creating contingency plans. These strategies enable managers and engineers to be prepared for any eventuality that may arise during project execution. Finally, the planning and design phase also facilitates communication with other key stakeholders, such as regulatory authorities and the local community. Transparency and clarity at this stage make it possible to anticipate and resolve conflicts of interest, obtain necessary permits and approvals, and establish trusting relationships with stakeholders. Collaboration between engineers and managers during this stage not only ensures that the project is technically and financially feasible, but also helps to create a more cohesive project tailored to the expectations of all stakeholders. This thorough planning allows for smoother project execution and significantly reduces the risk of encountering problems that could affect the project's viability, quality and sustainability in the future.

1.3 Resource Management and Budgeting

The management of financial and human resources is a critical task in the execution of infrastructure projects, as it ensures that each resource is used efficiently and sustainably throughout the various phases of the project. Financial and operational sustainability is highly dependent on the accurate planning and allocation of these resources, which allows the project team to meet its objectives within the established budget and time constraints, while avoiding waste and cost overruns that can compromise the project in the long run. (Meredith, J. R., & Mantel, S. J. (2017).

From a financial perspective, management is responsible for establishing a detailed budget that covers all project activities from initial planning through implementation and maintenance (Cleland & Ireland, 2013). This budget includes line items for each component, such as materials, labour, equipment, permits and other indirect costs. In addition, infrastructure projects often involve unforeseen costs, such as fluctuations in the price of materials or additional costs resulting from delays. Project managers must anticipate these potential costs and include a financial reserve in the budget to cover them without affecting the progress of the project. To do this, they use cost estimation and risk management techniques to help calculate the necessary investment more accurately. In addition, financial management is responsible payment scheduling and cash flow, ensuring that financial resources are available at the right time and in the right amount for each stage of the project. This is particularly important in infrastructure projects, where payment schedules for suppliers, contractors and other services must be aligned with the delivery times of each phase. Proper cash flow management avoids disruptions and ensures that there are no delays caused by lack of funds at key points in the project (Bent & Humphreys, 2009).On the other hand, human resource management is equally essential for the sustainable development of the project. In large-scale infrastructure projects, the team may range from labourers and technicians to engineers, architects, project managers, project managers, and other professionals. project and specialists in different areas. Management is responsible for coordinating and optimising the use of all these human resources, ensuring that each person is assigned to the tasks that best suit their skills and experience. This task allocation process must also consider operational efficiency, so that there is no duplication effort or downtime that can reduce the productivity of the team. (Chan, A. P. (2014). Engineering, for its part, plays a key role in optimising the use of materials and technologies within the established budgetary framework. One of the ways engineers achieve this is by selecting highly durable materials and advanced construction technologies that improve efficiency without significantly increasing costs. For example, they can choose construction techniques that reduce the amount of waste or speed up construction times, saving both financial resources and labour. In addition, engineers can employ modelling and simulation tools that allow them to predict the behaviour of materials and the final structure, minimising the possibility of costly errors (Project Management Institute, 2021).

The synergy between management and engineering in the use of resources is essential to adapt costs to the reality of the project, which translates into greater sustainability. To maximise the benefits of this collaboration, both areas work in constant communication and adjust strategies according to the specific needs that arise during the project. This collaboration is key to developing an infrastructure that is not only functional, but also economical and resource-efficient (Chan, A. P. (2014).

1.4 Quality Control and Compliance Regulatory Compliance

Quality control and regulatory compliance are fundamental pillars in the development of infrastructure projects, as they ensure that works are executed under the highest technical standards and in accordance with current laws and regulations. These aspects are critical not only to ensure the safety and durability of structures, but also to protect project stakeholders from legal risks and possible sanctions that could result from non-compliance. In this context, collaboration between engineers and managers is key to achieving optimum quality, from the design stage through to completion and handover.

From a technical perspective, engineers are responsible for ensuring that every aspect of the project meets established quality , which often defined by safety, efficiency and functionality regulations. This includes testing materials, verifying structural strength and complying with detailed specifications in plans and designs. In addition, engineers supervise the execution of works to ensure that procedures are carried out in accordance with recommended practices and that appropriate tools and technologies are used. Periodic audits and technical reviews are essential to detect potential errors at an early stage and correct them before they represent significant problems or put the integrity of the construction at risk. (Project Management Institute. (2021).

Quality control also involves the implementation of quality management systems, such as ISO 9001 standards in infrastructure projects. These standards provide a structured framework to help manage and optimise construction processes by establishing verification procedures and documentation of each stage. The application these systems not only facilitates quality control, but also allows for monitoring and continuous improvement, resulting in a final project that meets the highest technical and client satisfaction standards. (Project Management Institute. (2021).

From an administrative point of view, the responsibility for ensuring that the project complies with local and international regulations rests with the management team, who ensure that each stage of the project is in compliance with legal requirements. In infrastructure, the regulatory framework often covers a wide range of issues, including zoning laws, environmental standards, occupational safety regulations, and region-specific construction standards. Management is responsible for coordinating with regulatory authorities, obtaining the necessary permits and ensuring that all activities comply with legal provisions. This work is crucial to avoid penalties, fines and work stoppages that could result in financial losses and project delays (Kerzner, H. (2017).

Collaboration between engineering and management in quality control and compliance allows problems to be addressed comprehensively and effectively. Engineers provide the technical information needed to meet quality standards, while the management team ensures that all regulations and standards are taken into account. Together, they work to establish a compliance plan that covers both technical and legal requirements, creating a safe and working environment. This collaboration significantly reduces legal risks by ensuring that every aspect of the project has been considered in detail and that the necessary documentation and approvals are in place for implementation (Rodríguez, A., González, L., & Ramírez, M. (2019). In addition, compliance and quality control are not only limited to the construction phase, but also extend to the maintenance and operation of the infrastructure. Management is responsible for ensuring that the facilities continue to comply with current regulations and that periodic reviews are carried out in accordance with the recommendations of the engineers. This continuity is essential to keep the infrastructure in optimal condition, ensure the safety of users and extend the useful life of the project, which in turn contributes to the economic and social sustainability of the project (Kerzner, H. (2018). Ultimately, quality control and regulatory compliance are critical to the success and sustainability of any infrastructure project. Collaboration between engineering and management on these aspects not only ensures high quality infrastructure, but also protects everyone involved and benefits the community at large by providing a safe, functional and compliant project. (Schwaber, K., & Sutherland, J. (2017)

1.4.1Technological Innovations in Infrastructure

From an engineering perspective, the implementation of new technologies has led to significant advances in construction efficiency, quality and safety. One of the most transformative tools in this area is Building Information Modelling (BIM), which allows engineers to create accurate digital representations of projects before physical construction begins. This technology not only improves planning and coordination between different teams, but also makes it easier to identify potential problems in the early stages of the project. With BIM, it is possible to simulate the performance of the building over its lifetime, helping to optimise design and select materials that minimise environmental impact. This is especially valuable in large and complex projects, where coordination errors can lead to significant costs (Eastman et al., 2011). In addition, the incorporation of drones in the monitoring and assessment of construction sites has radically changed the way infrastructure projects are managed. Drones allow for quick and accurate aerial inspections, making it easier to monitor large areas and collect data in real time. This not only improves safety by reducing the need for workers to access hazardous areas, but also allows engineers to make informed decisions based on accurate and up-to-date data. Drones are capable of creating detailed topographic maps and 3D models, which are crucial for planning and design. Another significant innovation in construction is 3D printing, which has proven to be an effective tool for creating customised structural components quickly and efficiently. This technology reduces material waste and production costs, while allowing for greater flexibility in design. 3D printing can not only be used to create complex architectural elements, but also being explored in the construction of housing, bridges and other infrastructure. Through this technique, houses can be printed in a short period of time, which could offer quick solutions to the housing crisis in various parts of the world (Wu et al., 2016).

1.4.2Sustainable Infrastructure Practices

Parallel to these technological advances, project management plays a crucial role in integrating sustainable practices into modern infrastructure. Decisions on material selection, waste management and energy efficiency are now an integral part of project planning and implementation. The administration is working to establish policies that encourage the use of recyclable and sustainable materials, as well as construction practices that reduce environmental impact. This includes implementing environmental management systems that assess and minimise the ecological footprint of each project, ensuring compliance with environmental regulations and obtaining the necessary certifications, such as LEED (Leadership in Energy and Environmental Design) and BREEAM (Building Research Establishment Environmental Assessment Method). Sustainability is not only about using recyclable or responsibly sourced materials, but also about considering the full life cycle of products used in construction. This refers to assessing the environmental impact of materials from extraction to final disposal. Project managers are now increasingly focused on selecting products that are not only functional, but also have a lower carbon footprint. For example, the use of recycled concrete and the incorporation of bio-components in building materials, which can significantly reduce environmental impact, have been promoted (Giesekam et al., 2016). Management should also address the concept of passive design, which optimises the energy efficiency of buildings by carefully planning their orientation, using temperature-regulating materials and maximising natural light. The implementation of passive design strategies can reduce reliance on mechanical air conditioning systems, thereby reducing energy consumption and associated emissions. This approach, combined with active technologies such as solar panels and water recovery systems, can transform a building into a truly sustainable structure (Smith et al., 2017).

1.4.3Sustainability in Infrastructure Rehabilitation

It is important to note that sustainability in infrastructure is not limited to the construction of new buildings or structures. It also extends to the maintenance and rehabilitation of existing infrastructure. Management should consider sustainable renovation that extend the useful life of facilities without the need to demolish and rebuild, which is often resource intensive. Rehabilitation techniques include restoring original materials and retrofitting old systems, ensuring that infrastructure can meet current needs without compromising its integrity or generating excessive waste (Becerik-Gerber, B., Jazizadeh, F., Li, N., & Calis, G. (2012). For example, in the case of bridge rehabilitation, engineers can use advanced technologies to assess the existing structure, identifying areas that require strengthening without the need for complete reconstruction. This not only saves resources, but also preserves the history and character of the existing infrastructure. Material reuse and recycling practices in rehabilitation work are equally crucial to minimising environmental impact and reducing costs (Cleland, D. I., & Ireland, L. R. (2013).

1.4.4Interdisciplinary Collaboration

Collaboration between engineers and managers is essential to ensure that technological innovations and sustainable practices are effectively integrated into projects. This process involves a joint working approach to research and development of new solutions that respond to contemporary infrastructure challenges. For example, engineers can develop new construction techniques that incorporate sustainability criteria, while management ensures that these techniques are implemented within the established budgetary and regulatory framework.Knowledge sharing between these disciplines is essential to problems holistically. For example, engineers need to be aware of the financial implications of the technologies they propose, and managers need to be aware of the financial implications of the technologies they propose. informed about the latest innovations available in the engineering field. Together, they can create an environment where sustainability is prioritised, not just as a regulatory requirement, but as an essential practice that fosters innovation and efficiency in all aspects of the project.

1.4.5Sustainability as a Necessity Overriding Need

In the context of modern infrastructure, sustainability is no longer optional, but an imperative. As cities grow and the demand for infrastructure increases, it is essential to adopt approaches that minimise environmental impact and maximise resource efficiency. Integrating new technologies and sustainable practices into the planning, design and construction of infrastructure not only contributes to the creation of more resilient and future-proof environments, but also promotes sustainable economic development that benefits communities and the planet as a.Global pressure to combat climate change has led many countries to set ambitious targets for carbon emission reductions and renewable energy use. In this context, infrastructure projects should be aligned with these targets, promoting the construction of buildings, sustainable transport systems and utility networks that minimise waste and maximise the use of renewable resources.

1.4.6Long Term Implications

Finally, it is critical to consider the long-term implications of implementing sustainable technologies and responsible building practices. As the world's population continues to grow, the demand for infrastructure is expected to increase. Therefore, the planning and construction of these infrastructures must be done in a way that considers not only the current context, but also the impact they will have on future generations. This implies a precautionary approach that considers the adaptability of infrastructure in the face of climate change and the social and economic challenges that may arise.

1.5 Conclusion

In conclusion, engineering and management are two interdependent disciplines infrastructure construction. Their effective collaboration is fundamental to the success of any project, as it ensures that not only technical and quality requirements are met, but also that financial viability and long-term sustainability are achieved. Synergy between engineers and managers allows for more comprehensive and adaptive planning, able to anticipate and mitigate risks. This collaborative approach is especially important in an ever-changing global context, where environmental and social challenges demand innovative and responsible solutions. Furthermore, integrating sustainability into the design and delivery process not only optimises resources and minimises environmental impact, but also ensures that infrastructure is resilient and adapts to the future needs of communities (Project Management Institute [PMI]. (2017). The intersection of engineering and management, therefore, becomes an essential pillar for the development of infrastructures that not only respond to today's demands, but are also capable of enduring over time, offering tangible benefits for generations to come. This approach not only promotes equitable and sustainable development, but also fosters a culture of innovation and shared responsibility, fundamental to building a more prosperous and sustainable future for all.

CHAPTER 2
RESOURCE MANAGEMENT AND SUSTAINABILITY IN INFRASTRUCTURE

Resource management and sustainability are key elements in the development of infrastructure projects. In a world where urban growth and climate change present increasing challenges, the way in which resources are managed and sustainability is applied has become a determining factor for the success of any infrastructure project. Proper management of material, financial and human resources is essential to ensure not only the economic viability of the project, but also its positive impact on the environment and its ability to meet the needs of current and future generations. Proper management allows for an efficient use of resources, avoiding waste, reducing costs and minimising negative impacts on the environment. Material resource management involves thorough planning to determine the quantity, type and quality of materials required, as well as the logistics necessary to ensure their availability at the right time. Selecting sustainable and low environmental impact materials, such as those that are recycled or responsibly produced, contributes significantly to the overall sustainability of the project. In addition, factors such as durability and life cycle of materials must be considered to ensure that the infrastructure built is resilient and can withstand adverse conditions with minimal maintenance over time (Vanegas, 2019). Financial management also plays a crucial role in the sustainability of the project. Financial planning should include not only the direct costs of construction, but also the long-term costs associated with the maintenance and operation of the infrastructure. A sustainable financial approach involves identifying funding sources that support responsible practices, such as funds earmarked for green projects or public-private partnerships that promote sustainable development. In addition, efficient financial management helps to optimise the cost-benefit ratio, ensuring that every investment made translates into real infrastructure improvements and tangible community benefits. In terms of human resource management, it is essential to have a skilled team committed to the sustainability objectives of the project. Ongoing training in sustainable construction techniques, the use of new technologies and compliance with environmental regulations allow workers to perform their tasks more efficiently and aware of the impact they may have on the environment. In addition, a good working environment and the appropriate allocation of responsibilities contribute to greater productivity and, therefore, to a better use of available resources (Rodríguez, A., González, L., & Ramírez, M. (2019). The focus on sustainability is not limited to the construction stage; it must also be present in the operation and maintenance of infrastructure. This means designing projects that are energy efficient, use renewable energy sources and are prepared to adapt to changing conditions, such as rising temperatures or more frequent extreme weather events. Collaboration between engineers and managers is essential to ensure that these aspects are integrated from the design and planning phase, allowing the infrastructure to be functional and sustainable throughout its life cycle (Cortés, 2018). This chapter delves into how management and engineering work together to optimise resources and promote sustainable practices. Integrating sustainable practices into all phases of a project, from planning to operation, not only contributes to the preservation of the environment, but also ensures that the infrastructure meets today's standards of quality and resilience. Through effective management and interdisciplinary collaboration, it is possible to create infrastructure that not only meets the needs of today's society, but also protects resources and the well-being of future generations. (Occupational Safety and Health Administration [OSHA]. (2020).

2.1 Financial Resources Management

Financial management in infrastructure involves the planning, allocation and monitoring of the financial resources required for each phase of the project, from conception and design to implementation, operation and maintenance. It is one of the most critical functions in the management of infrastructure projects, as it ensures the availability of funds in the most efficient way. the right time, thus ensuring that the project moves forward without interruption and with an efficient allocation of financial resources (Arce-Ruiz, 2017). Financial planning is the first and most important step, as it allows to anticipate all costs involved and to establish a detailed budget covering all project activities. This planning includes not only construction costs, but also those associated with preliminary studies such as environmental assessment, feasibility studies and design. In addition, costs such as preventive and corrective maintenance, infrastructure operations after completion and eventual refurbishment should be considered. A detailed and well-planned budget can prevent most of the financial problems that often affect infrastructure projects, such as cost overruns and delays (Cortés, 2018). Financial resource allocation, on the other hand, is the activity that ensures that each area of the project has the necessary funds at the right time. This involves prioritising expenditure, ensuring that cash flow is sufficient, and efficiently managing income and expenditure. For complex infrastructure, this means forecasting payments to material suppliers, contractors, subcontractors and service providers, and following up to ensure that all financial obligations are met in a timely manner. Management must be able to anticipate any deviations in cash flow and have strategies in place to mitigate these deviations, such as seeking additional funding or rescheduling payments. (Lock, D. (2020). Constant monitoring of the financial flow is equally essential to ensure that the project stays within budget. Infrastructure projects are often large and extend over months or even years, so it is common for changes or unforeseen events to affect the initial cost estimate. This is where collaboration between managers and engineers is crucial. Engineers have a thorough understanding of the technical needs of the project and can identify if any changes to the design or materials are necessary. However, any modifications must be evaluated from a financial perspective by the managers, who will determine if the change is feasible within the budget or if it is necessary to make a change in the design. additional adjustments. This joint assessment process allows for minimising financial risks and making informed decisions that benefit both the technical quality of the project and its economic viability.

Another important aspect of financial management in infrastructure is the identification of financing sources. Infrastructure projects often require large sums of money, and are therefore commonly financed through a variety of sources, including public finance, private finance, bank loans, grants, public-private partnerships (PPPs) and, in some cases, through multilateral funding agencies. Financial management must be able to identify the best mix of funding sources based on project characteristics, cost of money, repayment terms and timing. The right approach to raising finance can significantly reduce the financial burden of the project and improve its profitability (International Organization for Standardization [ISO] (2015b).

Financial risk management is also an integral component of financial management in infrastructure. Projects of this nature are exposed to multiple risks, such as fluctuations in the cost of materials, changes in interest rates, delays that entail additional costs, and economic or political risks that may affect financing. Managers should conduct a detailed risk assessment and develop strategies to mitigate these risks. This could include using fixed-price contracts to avoid cost fluctuations, taking out insurance to cover unexpected events, and creating contingency funds within the budget to cover unforeseen expenses

(Walker, A., & Rowlinson, S. (2008).The collaboration between management and engineering in financial management also includes optimising the use of economic resources, which can involve the implementation of more efficient construction technologies and techniques that, although they may initially appear more costly, lead to significant savings in the long term. For example, the adoption of methodologies such as Lean Construction can reduce waste and improve efficiency in the use of materials and , which directly contributes to the optimisation project costs. Engineers can propose technological innovations that will optimise The project's performance, while managers assess the economic viability of such proposals and ensure that they are implemented effectively and within budget (Lingard, H., & Rowlinson, S. (2005).

Finally, transparency and accountability in financial management are essential for project success. Infrastructure is often financed with public resources or with funds that involve a high degree of accountability to various stakeholders, including governments, financial institutions and the community. Maintaining accurate financial records, conducting regular audits and providing clear reporting on the financial status of the project are key to building trust and ensuring that the project is completed successfully and without financial setbacks (Chan, A. P. (2014).

2.2 Human Resources Management

Human resources are critical to the success of any project, as people are responsible for executing tasks, making decisions, and ensuring that every aspect of the project is carried out effectively and efficiently. In the context of infrastructure projects, human resource management is especially critical due to the complexity and magnitude of these projects, which involve multiple phases, technologies, and a rigorous focus on safety and quality of work (Environmental

Protection Agency [EPA]. (2016). Human resource management in infrastructure projects encompasses several essential activities, such as recruitment, training, motivation and the allocation of tasks to the right workers for each activity. Recruitment is the first stage and focuses on selecting individuals who possess the technical skills and experience necessary to contribute to the success of the project. This requires a thorough selection process to assess candidates not only in terms of technical skills, but in terms of their ability to work as part of a team, adapt to change and handle stressful situations, all of which are key characteristics in infrastructure projects (Franco Reina, 2022). Once the team is in place, training becomes a key component to ensure efficiency and safety in the execution of tasks. Modern infrastructure often involves the use of advanced technologies, heavy machinery and innovative methodologies that require specific skills. Management must ensure that all workers receive the necessary training to properly use the required tools and equipment, as well as to meet project quality standards. In addition, training in safety practices is essential to minimise risks and ensure that all employees understand the necessary measures to protect their integrity and that of their colleagues (Vanegas, 2019).

Motivation and retention of talent are also crucial aspects of human resource management. Keeping a motivated and committed team can make the difference between the success and failure of a project. Management should implement policies that promote employee well-being, such as creating a safe and positive work environment, providing incentives for meeting goals, and recognising outstanding effort and performance. Motivating employees not only improves productivity, but also reduces staff turnover, which is especially beneficial in long-term projects where continuity and accumulated experience are of great value (Ballard, G., & Howell, G. (2003). Task allocation is another key responsibility in human resource management. Each worker must be assigned to activities that best match his or her skills and experience, which maximises productivity and minimises errors. This involves continuous coordination between project managers and engineers to clearly define the technical profiles required for each stage of the process and to ensure that assigned personnel possess the appropriate competencies. In infrastructure projects, the correct allocation of human resources is crucial, as the work performed by each individual directly influences the quality and safety of the infrastructure being built. The management of labour and safety regulations is also a key part of human resource management. Infrastructure projects are often subject to strict regulations in terms of occupational safety, both because of the nature of the work involved, which is often carried out in hazardous conditions, and because of the machinery and materials used. Management must ensure that all legal regulations are complied and that safety policies are implemented to protect workers. This includes providing personal protective equipment (PPE), conducting regular safety inspections, implementing safety training programmes, and developing emergency response protocols. Complying with these regulations not only avoids penalties and work stoppages, but also contributes to the creation of a safe working environment and the reduction of workplace accidents. For their part, engineers work hand in hand with managers to define the technical profiles required for each phase of the project. This includes determining what specific skills are needed for the installation of complex systems, the use of specialised machinery, or the implementation of new construction technologies. Engineers provide the technical knowledge and can provide guidance on the skills and competencies required, while managers identify and recruit the most appropriate personnel. This collaboration ensures that each task is performed by a skilled worker, reducing the risk of errors and improving project efficiency.

In addition, communication and teamwork are fundamental aspects of human resources management in infrastructure projects. Coordination between the different teams - engineers, workers, managers, suppliers and contractors - requires effective management that facilitates communication between all the actors involved. This communication must be fluid to ensure that everyone is aware of the project's progress, changes that may arise and everyone's responsibilities. Managers must foster a collaborative working environment, where each team member feels valued and able to contribute to the success of the project. An increasingly relevant aspect of human resource management is diversity and inclusion in infrastructure projects. Organisations that promote diversity in their teams find multiple benefits, such as the emergence of new ideas, a greater problem-solving skills and an enriched work environment. Human resource management should strive to create diverse teams that include people of different backgrounds, genders and experiences, and ensure that everyone has equal opportunities for professional development and growth. Finally, successful human resource management in infrastructure also involves long-term planning that includes the development of staff competencies. In projects that extend over several years, it is important to invest in the skills development and career development of employees. Continuous training programmes, promotion opportunities and a culture of constant learning are elements that help to maintain a highly qualified and motivated team, ready to face the challenges of the project at every stage.

2.3 Materials and Technology Management

The management of materials and technologies plays a crucial role in the construction of sustainable infrastructure, as these decisions directly affect both the environmental impact of the project and its efficiency and viability. Proper selection of materials and use of innovative technologies reduces waste, improves construction quality and contributes to the sustainability of the project. Effective collaboration between engineers, who specify the most appropriate materials and technologies, and managers, who manage their logistics and costs, is essential to ensure that the project's quality, sustainability and efficiency objectives are met. In terms of building materials, the selection should be based on criteria of quality, durability and sustainability. High-quality and durable materials are essential to ensure the longevity of infrastructure, which is an essential component of sustainability. Choosing materials that withstand weathering and wear over time reduces the need for frequent repairs and renovations, helping to minimise resource use and waste generation over the life cycle of the project. For example, materials such as high-strength steel and concrete with special admixtures can significantly increase the durability of structures. In addition to durability, sustainability means selecting materials with a low environmental impact. This means opting for those that have a lower carbon footprint during production, transport and disposal. The use of recycled and recyclable materials has become an increasingly common and accepted practice in modern construction. For example, recycled concrete, recycled steel and certified wood from responsibly managed forests are materials that reduce the exploitation of natural resources and the environmental impact of the project. Engineers play an important role in evaluating the properties of these materials to ensure that they meet the required quality and safety standards. The life cycle of materials is another critical factor for engineers to consider. This involves assessing the environmental impacts of materials during all stages of their life cycle, from the extraction of raw materials to their eventual disposal or recycling. This approach allows for the identification of materials that are not only more sustainable during the construction phase, but also contribute to reducing impacts. Engineers must keep abreast of new research and developments in sustainable materials to propose the best available options, while managers must assess the economic viability of these proposals and seek a balance between cost, quality and sustainability (Cleland, D. I., & Ireland, L. R. (2013). The management of logistics and availability of materials is the responsibility of the administration, and is a crucial component in ensuring efficient project execution. Logistics involves coordinating the supply of materials so that they are available at the right time and in the right quantity, avoiding delays in construction. This requires detailed planning to synchronise the delivery of materials with the construction schedule, ensuring that each phase of the project has the necessary resources without creating unnecessary storage, which could lead to loss or deterioration. Management must also manage the relationship with suppliers, selecting those who can guarantee the quality of materials and their timely delivery. Cost control of materials is another fundamental aspect that falls under the responsibility of management. Materials represent a large part of the total cost of any project. infrastructure costs, so proper cost management is crucial to keep the project within budget. Managers must continuously keep track of costs, considering not only the purchase price, but also the costs associated with transporting, storing and handling materials. In addition, they must be prepared for fluctuations in the price of materials, which can be affected by factors such as market availability, economic conditions and fuel costs. In these cases, management must be able to seek viable alternatives or renegotiate contracts to mitigate the financial impact on the project.

In terms of construction technologies, technological innovation plays a crucial role in improving the efficiency and sustainability of infrastructure projects. Modern technologies make it possible not only to optimise construction processes, but also to reduce the environmental impact. For example, the use of Building Information Modelling (BIM) allows engineers and managers to work with three-dimensional digital models of the project, which facilitates planning and informed decision-making. With BIM, engineers can evaluate different design options and materials, simulating their behaviour and impact before construction begins, helps to minimise waste and optimise the use of resources.

Another technology that is revolutionising construction is 3D printing, which allows structural components to be manufactured accurately and efficiently. This technology not only reduces material waste, but also shortens construction times and allows the creation of complex designs that would be difficult to achieve with traditional methods. 3D printing can be used both for the fabrication of structural elements and for the production of decorative or functional parts, and its adoption is starting to gain traction in the construction industry.

In addition, the adoption of technologies such as prefabrication and prefabricated modules has also proven to be effective in improving the efficiency and sustainability of infrastructure projects. Prefabrication involves the construction of infrastructure elements in a controlled environment, off-site, and then transported and assembled at their final location. This technique significantly reduces the time it takes to construction and improves the quality of work, as elements are manufactured under optimal conditions and are less prone to errors. Prefabrication also minimises the impact on the construction site, reducing waste generation and emissions associated with transport and the use of heavy machinery on site.

Collaboration between engineers and managers is essential to maximise the benefits of materials and technology management. Engineers must identify the most appropriate materials and technologies based on project objectives, ensuring that they meet technical and sustainability standards. At the same time, managers must ensure that the selection of materials and technologies is economically viable and that their implementation is logistically and cost efficient. This collaboration helps to optimise the use of resources, reduce waste, ensure project quality and meet sustainability objectives.

2.4 Approaches Sustainable

Sustainability is a critical aspect of modern infrastructure construction, and its relevance has increased exponentially in recent years due to the challenges of climate change, natural resource depletion and the need to build resilient urban environments. Sustainable infrastructure not only seeks to minimise environmental impact during construction, but is also designed to operate efficiently throughout its life cycle, optimising the use of energy, water and other resources, and contributing to the quality of life of communities.

One of the key components of sustainability is the use of recycled and sustainable materials. Choosing materials that have been recycled or can be recycled at the end of their useful life helps to reduce raw material consumption and waste generation. Materials such as recycled concrete, reused steel and wood from certified sources are examples of alternatives that can contribute significantly to reducing environmental impact. In addition to using recycled materials, engineers should also consider the carbon footprint of the materials they select, opting for those that emit fewer greenhouse gases during production and transport. Reducing the environmental impact from the selection of materials is an essential strategy for building more sustainable infrastructure.

Energy efficiency is another fundamental pillar of sustainability in modern infrastructure. In the design of efficient structures, engineers are responsible for incorporating measures that reduce energy consumption, both during construction and operation of the infrastructure. This includes the design of daylighting systems, the integration of efficient ventilation systems, the use of adequate thermal insulation and the installation of renewable energy technologies such as solar panels or wind turbines. Energy efficiency not only contributes to reducing environmental impact, but also translates into long-term economic savings, which is beneficial both infrastructure owners and end-users.

Designing efficient infrastructure also requires careful planning to reduce carbon emissions. This includes selecting construction methods that minimise the generation of emissions, such as the use of efficient machinery and planning processes to reduce the amount of time and energy required. For example, the adoption of modular construction technologies can significantly reduce energy consumption during construction by allowing components to be manufactured in a controlled environment and then assembled on site, minimising the use of heavy machinery and the emission of pollutants.

Proper management of waste generated during construction is another key sustainable practice. In infrastructure projects, the amount of waste generated can be considerable, and how that waste is managed has a direct impact on the environment. Implementing strategies to reduce, reuse and recycle surplus materials is critical to minimising the impact. This involves the development of a waste management plan that includes the recycling of debris, the reuse of materials and proper disposal of waste that cannot be recovered. In addition, proper planning and the adoption of more accurate construction techniques can help to significantly reduce the amount of waste generated. In this context, engineers play a key role in the design of sustainable structures and the implementation of green technologies. During the design phase, engineers must consider factors such as resource efficiency, integration of renewable energy systems and minimisation of environmental impact throughout the life cycle of the project. Design must be adaptive, anticipating the ability of infrastructure to evolve with future needs, which is particularly relevant in a context of increasing urbanisation and climate change. Engineers also have a responsibility to innovate and explore new technologies that can contribute to the sustainability of projects, from greener materials to construction processes that minimise the impact on the environment. On the other hand, project managers are instrumental in sustainable policies and strategies throughout the infrastructure lifecycle. Managers must ensure that decisions made during the planning, construction and operation of infrastructure are aligned with the sustainability objectives of the project. This includes ensuring compliance with local and international environmental regulations, seeking environmental certifications such as LEED or BREEAM, and managing resources efficiently to minimise economic and environmental impact. In addition, management must evaluate the cost-benefit of sustainable technologies and practices, always seeking a balance between economic viability and sustainability. An important aspect of integrating sustainable practices is the promotion of a culture of sustainability among all actors involved in the project, from engineers and constructors to end users. Awareness raising and staff training are essential to ensure that everyone understands the importance of sustainability and how they can contribute to it in their daily tasks. Managers play a key role in implementing training programmes and creating incentives to promote sustainable practices among project staff. Sustainability during the operation and maintenance phase of infrastructure is another critical factor. It is not enough to design and build sustainable infrastructure; it is also necessary to ensure that they are operated and maintained efficiently so that they continue to meet their sustainability objectives. This involves implementing monitoring and control systems to assess energy and water consumption, as well as identifying areas where efficiency can be improved. Management should plan preventive and predictive maintenance activities to prolong the life of infrastructure and avoid costly and disruptive repairs. Adaptability and resilience of infrastructure are also essential components of modern sustainability. As climates change and cities grow, infrastructure must be able to adapt to new conditions and demands. This includes designing structures that can withstand extreme weather conditions, such as floods, heat waves or earthquakes, and that are flexible to allow for future modifications and expansion. Resilience has become a pillar of sustainability, as it ensures that infrastructure can remain functional and safe even in the face of unexpected events.

2.5 Partnership for Sustainability

Collaboration between management and engineering is essential for the efficient management of resources and for ensuring sustainability in infrastructure projects. This interdisciplinary cooperation not only facilitates effective planning and rational use of resources, but also ensures that sustainable practices are adopted and maintained at every stage of the project, from conception to operation and maintenance. The integration of knowledge and skills from both disciplines allows project challenges to addressed holistically, considering technical, financial, human and environmental aspects simultaneously.

Joint planning between management and engineering is the basis for the success of any infrastructure project. Engineers bring technical expertise to design and construction, ensuring that the solutions adopted are technically feasible and meet quality standards. At the same time, managers bring a financial and strategic perspective, ensuring that technical decisions are economically viable and sustainable over time. The collaboration between the two areas allows that clear and achievable objectives are defined, and that realistic timelines and well-founded budgets are established. This comprehensive planning not only facilitates the optimal use of financial, human and material resources, but also helps to anticipate and mitigate potential risks that may affect the development of the project. Financial resource management is one of the areas where collaboration between management and engineering is most evident and necessary. Management is responsible for the planning and allocation of financial resources, while engineers provide the information needed to assess the costs associated with each technical aspect of the project. Together, they can identify areas where waste can be reduced, optimise the use of materials and select technologies that are cost-effective and sustainable. This collaboration also allows them to assess the feasibility of adopting more advanced technologies or sustainable materials that, while they may have a higher initial cost, bring significant benefits in terms of long-term efficiency and sustainability. In terms of human resource management, collaboration between management and engineering is equally crucial. Managers are responsible for recruiting and assigning personnel, ensuring that sufficient skilled workers are available for each phase of the project. Engineers define the necessary technical profiles and provide technical training to ensure that each team member is prepared to execute their tasks efficiently and safely. Joint human resource planning allows responsibilities to be allocated appropriately, avoiding duplication of effort and ensuring that each worker is involved in activities that align with their skills and experience. This not only improves operational efficiency, but also contributes to staff motivation and satisfaction.

Material resource management also benefits greatly from collaboration between management and engineering. Engineers are responsible for specifying the materials needed for the project, assessing their durability, quality and sustainability. Managers, on the other hand, are responsible for ensuring the logistics, availability and cost of these materials. Working together, they can look for alternatives that are both technically and economically feasible, ensuring that high quality materials are selected whichare durable and have a low environmental impact. Management is also responsible for coordinating the supply of materials, ensuring that they are available at the right time, which is essential to avoid delays and cost overruns during construction. The integration of sustainable practices in all phases of the project is one of the main benefits of the collaboration between management and engineering. Engineers are responsible for identifying and proposing technical solutions to improve the sustainability of the project, such as the use of recycled materials, the implementation of energy efficiency systems and the minimisation of waste during construction. Managers are responsible for implementing policies and strategies to ensure that these practices are maintained and effectively applied during all phases of the project. This includes complying with environmental regulations, pursuing sustainability certifications, and coordinating with the various stakeholders involved in the project to ensure an aligned approach to sustainability. Collaboration between management and engineering is also essential to mitigate risks and ensure project resilience. During the planning phase, engineers and managers work together to identify potential risks, such as technical issues, financial constraints, environmental impacts or safety risks. Once identified, mitigation strategies are developed to reduce the likelihood of these risks materialising and minimise their impact should they occur. The collaboration between the two disciplines allows risks to be addressed in a comprehensive manner, considering not only the technical aspects of the project, but also its economic, social and environmental implications. Another advantage of management-engineering collaboration is the ability to maximise the viability of the project. Sustainability is not only limited building infrastructure that minimises environmental impact during the construction phase; it also involves designing and managing infrastructure that is efficient and sustainable throughout its life cycle. Engineers, in collaboration with managers, must plan to ensure that the infrastructure is easy to operate and maintain, has low operating costs and can adapt to future changes, such as the growth in demand or climate change. This long-term vision allows building resilient infrastructure that will remain functional and useful for future generations, maximising its value and reducing its environmental impact over time.

2.6 Conclusion

Resource management and sustainability are fundamental pillars in the construction of efficient and resilient infrastructures. These two elements, when properly integrated at every stage of a project, create structures that not only meet the needs of today, but are also prepared to meet the challenges of the future. Effective collaboration between managers and engineers in the planning and execution of projects ensures that financial, human and material resources are optimised while reducing environmental impact, thus ensuring long-term success and sustainability. In a context of increasingly limited resources and growing demands for sustainability, it is imperative that management and engineering work together in an aligned manner. This collaboration is key to developing innovative solutions that maximise project efficiency and quality, minimise waste and promote infrastructure resilience. The integration of new technologies, sustainable practices and effective management strategies enables the creation of infrastructure that not only adapts to changing needs, but also actively contributes to the improvement of the environment and the well-being of communities. In short, the synergy between management and engineering is essential to achieve sustainable development for the benefit of present and future generations.

CHAPTER 3
IMPLEMENTATION AND MONITORING IN INFRASTRUCTURE PROJECTS

2.7 Approaches to the Efficient Implementation of Infrastructure Projects

The efficient implementation of infrastructure projects requires a comprehensive approach that includes technical, financial, logistical and environmental aspects, as well as detailed planning and constant monitoring of project progress. This approach must include risk assessment, optimal resource allocation and the flexibility to adapt to unforeseen changes that may arise during implementation. Project management techniques, such as Lean Construction methodology and the use of Building Information Modelling (BIM), can improve efficiency, reduce waste and optimise resources. Lean Construction seeks to maximise value for the client by eliminating activities that do not add value, while BIM facilitates comprehensive visualisation of the project and coordination between the various stakeholders. Coordination between the different teams involved and clear and effective communication are essential to ensure that all parties are working towards the same goals, thus minimising errors, conflicts and delays. To achieve efficient execution, it is essential to establish an open communication system that allows for continuous feedback between teams, fostering a collaborative environment. In addition, the integration of digital tools, such as collaborative platforms, project management software and real-time monitoring systems, facilitates communication and ensures that information is available to all stakeholders in a timely manner, enabling more agile and informed decision-making (Kerzner, 2018). Another key aspect of efficient implementation is schedule management. Planning should include specific milestones that allow progress to be assessed and adjustments to be made where necessary, ensuring that deadlines are met and available resources are optimised. A well-structured schedule should be flexible enough to adapt to contingencies, but also detailed enough to ensure that each activity has a defined start and end time. The ability to identify critical project paths allows managers and engineers to focus efforts on those tasks that, if not completed on time, could delay the overall project.

Quality management is equally important for the efficient execution of infrastructure projects. Engineers and managers must work together to establish quality standards from the outset and ensure that all materials and processes meet these standards. Implementing quality controls at each phase of the project allows problems to be identified before they become major obstacles, thus ensuring the delivery of infrastructure that meets technical requirements and stakeholder expectations. In addition, quality execution contributes to project sustainability, as well-built infrastructure requires fewer repairs and maintenance in the long term (Schwaber & Sutherland, 2017).

Resource allocation also plays a key role in efficient execution. The optimal allocation of labour, materials, machinery and equipment is necessary to ensure that each phase of the project has the necessary resources at the right time. Detailed planning must take into account material quantities, delivery times and equipment availability so that there are no interruptions that can delay execution. The use of resource planning systems, such as ERP (Enterprise Resource Planning), facilitates the management of these aspects, allowing better control and responsiveness to changing project needs.Finally, it is important to consider cost management during project implementation. A comprehensive approach to cost management involves not only meeting the budget, but also continuously looking for ways to reduce costs without compromising the quality of the project. Implementing Earned Value Management (EVM) techniques helps managers measure project performance in terms of cost and time, providing a clear view of whether the project is in line with financial objectives. In this way, engineers and managers can make informed and proactive decisions to correct any deviations and ensure project efficiency.

3.2 Ongoing Monitoring and Evaluation Strategies

Ongoing monitoring and evaluation are essential to ensure that the project remains aligned with the objectives set in terms of time, cost and quality. Monitoring strategies include the use of key performance indicators (KPIs) to assess the progress and quality of the project at each stage. Technological tools such as BIM, drones and project management systems also contribute to more accurate and efficient monitoring, facilitating informed decision making and enabling rapid adjustments when necessary (PMI, 2017). The use of key performance indicators (KPIs) allows the evaluation of key aspects of the project, such as physical progress, schedule adherence, quality control, job site safety and environmental impact. These indicators provide a quantitative view of project performance, helping managers and engineers to identify problem areas and implement corrective measures in a timely manner. KPIs should be defined during the planning stage and should be reviewed and updated as necessary to reflect changes in project objectives or external conditions (Lock, 2020).

The use of drones has revolutionised the way monitoring is carried out in infrastructure projects. These devices allow for detailed aerial inspections, capturing high-resolution images and videos that provide a comprehensive view of construction progress. Drones are especially useful in hard-to-reach or sites, where manual inspections would be complicated and time-consuming. The images obtained can be used to compare the current state of the construction site with plans and digital models, identifying discrepancies and helping to resolve problems before they become significant obstacles (ISO, 2015b). IoT (Internet of Things) sensor technology also plays an important role in the continuous monitoring of infrastructure projects. Sensors installed in different parts of the construction site can provide real-time data on environmental conditions, such as temperature and humidity, as well as on the performance of equipment and the stability of structures. This data allows engineers to make decisions based on accurate and up-to-date information, improving the safety, efficiency and quality of the project. The integration of these technologies into a centralised management platform allows for more comprehensive project control, facilitating the early identification of risks and the implementation of preventive measures.

3.3 Environmental Impact during Implementation

During the implementation of infrastructure projects, it is crucial to minimise the environmental impact associated with construction activities. This involves proper waste management, reducing emissions, protecting natural resources and good environmental practices. Collaboration between engineers, managers and environmental specialists can identify potential risks and mitigate impacts, promoting responsible and environmentally friendly construction. (International Organization for Standardization [ISO]. (2015a).

Waste management is a key aspect in reducing the environmental impact during the implementation of a project. Separating and recycling construction materials, such as concrete, metal and wood, not only helps to minimise the amount of waste sent to landfill, but also contributes to the circular economy by reintroducing materials back into the production chain. In addition, it is important to implement practices to minimise waste generation from the outset, such as using more precise construction techniques and properly planning the quantities of materials needed (Becerik-Gerber et al., 2012).

Reducing emissions of greenhouse gases and other pollutants is another essential component of mitigating environmental impact. This can be achieved through the use of fuel-efficient machinery, the adoption of electric or hybrid technologies, and the optimisation of construction operations to minimise the time spent using heavy machinery. In addition, the implementation of efficient transport plans that reduce unnecessary travel and promote carpooling also contributes to emissions reductions (Eastman et al., 2011).

Protecting natural resources during construction involves minimising the impact on local ecosystems. This may include protection of sensitive areas such as water bodies and habitats of protected species, as well as reforestation of areas affected by construction. Environmental impact assessment, carried out prior to the start of the project, is essential to identify potential risks and establish appropriate mitigation measures. During implementation, continuous monitoring is essential to ensure that the established measures are adhered to and that any unforeseen impacts are addressed immediately (Ahmed et al., 2019).

3.4 Occupational Health and Safety

Occupational health and safety is a key issue in the implementation of infrastructure projects due to the nature of the activities involved, which often present risks to workers. It is essential to implement a health and safety management system that includes risk identification, staff training, the use of personal protective equipment (PPE) and continuous monitoring of compliance

with regulations. Preventing accidents and promoting a safe working environment not only protects workers' health, but also contributes to project efficiency and quality (EPA, 2016).
Risk identification is the first step in ensuring safety on the construction site. This involves conducting detailed risk assessments prior to the start of the project and maintaining an ongoing assessment throughout implementation. Risks can include falls, machinery accidents, exposure to hazardous substances, among others. It is essential that risks are proactively identified and preventative measures are put in place to mitigate them, such as installing guardrails, marking hazardous areas and implementing safe machinery handling procedures.

Staff training is crucial to ensure that workers are aware of the risks to which they are exposed and know the best practices to prevent accidents. Training programmes should include emergency drills, training in the correct use equipment and tools, and the promotion of a safety culture in the workplace. . Training should be continuous and tailored to the specific needs of the project, ensuring that all workers, including new employees and subcontractors, receive adequate training before starting their tasks.The proper use of personal protective equipment (PPE) is another essential component of protecting workers. Managers should ensure that all workers have access to the necessary PPE, such as hard hats, gloves, reflective waistcoats, safety harnesses and protective eyewear, and that this equipment is used correctly at all times. In addition, it is important to conduct regular inspections to check the condition of PPE and replace any equipment that is damaged or worn, thus ensuring that workers are always protected (UNEP, 2017).Continuous monitoring of compliance with safety regulations is critical to maintaining a safe working environment. This includes conducting regular safety audits, constantly monitoring construction site activities and implementing corrective actions when non-compliances are identified. Vigilance also involves fostering a safety culture in which all workers feel responsible for their own safety and of their colleagues, encouraging open communication about potential risks and participation in safety improvement initiatives.

Promoting a safe and healthy work environment also has a positive impact on project productivity. A safe environment reduces the incidence of accidents and injuries, which in turn reduces work stoppages and the costs associated with medical treatment and workers' compensation. In addition, a safe and well-managed work environment improves worker morale, which contributes to greater motivation and efficiency in performing their tasks.

3.5 Risk and Contingency Management

Risk management is an essential component in the implementation of infrastructure projects, as these projects often face a variety of technical, financial, environmental and social risks. The early identification of risks and the elaboration of contingency plans anticipate potential problems and minimise their impact on the project. Collaboration between managers and engineers is crucial to comprehensively assess risks and develop effective mitigation strategies, ensuring the continuity and success of the project.

Risk identification should be carried out from the earliest stages of the project and be a continuous process throughout implementation. Technical risks may include design failures, problems with the quality of materials or difficulties during construction. Financial risks, on the other hand, may involve cost overruns due to fluctuations in material prices or delays that increase costs. Environmental risks include potential damage to ecosystems, while social risks include issues such as opposition from local communities or labour disputes. To identify these risks, it is essential to conduct comprehensive assessments and involve all stakeholders, from engineers to local authorities and the community (OSHA, 2020).

Once risks have been identified, mitigation plans should be developed to reduce the likelihood of these risks materialising and to minimise their impact if they do occur. These plans may include actions such as using alternative materials in case of supply problems, implementing safer construction processes to avoid accidents, or developing financial strategies to secure additional funds in case of cost overruns. The implementation of these measures should be constantly monitored, and mitigation plans should be reviewed and updated as necessary to adapt to changes in the project or the environment.Contingency plans are another fundamental tool in risk management. These plans set out the actions to be taken when a risk materialises, with the aim of minimising its impact and ensuring that the project can continue. For example, a contingency plan for a materials supply problem might include prior arrangements with alternative suppliers, while a plan for an environmental emergency might include the mobilisation of specialised teams to mitigate damage. Contingency plans should be clear, detailed and known to all members, so that they can be implemented quickly and effectively when necessary. Residual risk assessment is an important step after implementing mitigation measures. This process involves analysing the risks that remain after preventive measures have been implemented and determining whether these risks are acceptable or whether additional measures are required. Residual risk assessment allows managers and engineers to have a clear view of the level of risk exposure and to make informed decisions on how to proceed. Documentation of risks and of mitigation and contingency strategies is essential to ensure transparency and efficiency in risk management. All identified risks, together with the measures taken to mitigate them, should be recorded and regularly updated. This documentation serves as a reference for the project team and also for future projects, as it allows learning from past experiences and improving risk management capacity. In addition, organisational culture plays a crucial role risk management. Promoting a culture of open and proactive communication about risks helps to identify potential problems before they become major obstacles. Workers, engineers and managers should feel empowered to report risks and suggest solutions, without fear of retaliation. This culture of safety and prevention fosters an environment in which all team members are committed to identifying and mitigating risks.

Technology also plays an important role in risk management. Tools such as Building Information Modelling (BIM) and software-based risk management systems make it possible to visualise potential risks and assess the impact of different scenarios. BIM, for example, facilitates the detection of potential design conflicts before construction activities begin, while software systems allow for continuous risk monitoring and the generation of reports to assist in decision making.

CONCLUSION

Implementation and monitoring in infrastructure projects require careful management that integrates efficiency in execution, use of advanced technologies and sustainability. Risk management and continuous evaluation ensure on-target development, while the final review provides valuable insights for future initiatives.

REFERENCE

Ahmed, S., Farooqui, R., & Saqib, M. (2019). Drones and construction: The rise of the machines. International Journal of Construction Education and Research, 15(4), 325- 344.

Ballard, G., & Howell, G. (2003). Lean project management. Journal of Building and Environment, 6(2), 115-125.

Becerik-Gerber, B., Jazizadeh, F., Li, N., & Calis, G. (2012). Application areas and data requirements for BIM-enabled facilities management. Journal of Construction Engineering and Management, 138(3), 431-442.

Bent, J. A., & Humphreys, K. K. (2009). Effective project management through the application of cost and schedule control. CRC Press.

Chan, A. P. (2014). Construction project management handbook. Routledge.

Cleland, D. I., & Ireland, L. R. (2013). Project management: Strategic design and implementation. McGraw-Hill.

Eastman, C., Teicholz, P., Sacks, R., & Liston, K. (2011). BIM Handbook: A Guide to Building Information Modeling. Wiley.

Environmental Protection Agency [EPA] (2016). Construction and demolition debris recycling. https://www.epa.gov/recycle

Project Management Institute (2021). Guide to Project Management Body of Knowledge (PMBOK Guide). Project Management Institute.

International Organization for Standardization [ISO] (2015a). ISO 9001: Quality management systems. ISO.

International Organization for Standardization [ISO] (2015b). ISO 14001:

Environmental management systems. ISO.

Kerzner, H. (2017). Project management: A systems approach to planning, scheduling and control. John Wiley & Sons.

Kerzner, H. (2018). Project Management: A Systems Approach to Planning, Scheduling, and Controlling. Wiley.

Koskela, L. (1992). Application of the new production philosophy to construction. Stanford University.

Lingard, H., & Rowlinson, S. (2005). Occupational Health and Safety in Construction Project Management. Routledge.

Lock, D. (2020). Project Management. Gower.

Meredith, J. R., & Mantel, S. J. (2017). Project Management: A Managerial Approach. Wiley.

Occupational Safety and Health Administration [OSHA] (2020). Construction industry standards. https://www.osha.gov

Project Management Institute [PMI] (2017). A Guide to the Project Management Body of Knowledge (PMBOK Guide). PMI.

Rodríguez, A., González, L., & Ramírez, M. (2019). Sustainable infrastructure development: The role of engineering and project management. Journal of Infrastructure Systems, 25(1), 1-12.

Schwaber, K., & Sutherland, J. (2017). The Scrum Guide.

Smith, P. G., & Reinertsen, D. G. (2014). Developing products in half the time: New rules, new tools. John Wiley & Sons.

Walker, A., & Rowlinson, S. (2008). Procurement systems: A cross-sector project management perspective. Routledge.

AUTHORS' MINI-CURRICULUM VITAE

Alejandro Enrique Nesci Montalvan: Business Administration student with a focus on organisation and efficient management. He enjoys working in a team and considers himself a great thinker.

Ricardo José Baldiz6n L6pez: Civil Engineering student with a strong work ethic and teamwork skills. With a particular interest in the development and sustainability of infrastructure projects.

Carlos Alexander Mendoza Jacomino: Doctor of Science, full researcher and academic at the American University (UAM).

Printed by Books on Demand GmbH, Norderstedt / Germany